DC motor control using Real-Time Linux

Asim Zaman
Shams Ullah

DC motor control using Real-Time Linux

The control is in your finger tips

LAP LAMBERT Academic Publishing

Impressum / Imprint
Bibliografische Information der Deutschen Nationalbibliothek: Die Deutsche Nationalbibliothek verzeichnet diese Publikation in der Deutschen Nationalbibliografie; detaillierte bibliografische Daten sind im Internet über http://dnb.d-nb.de abrufbar. Alle in diesem Buch genannten Marken und Produktnamen unterliegen warenzeichen-, marken- oder patentrechtlichem Schutz bzw. sind Warenzeichen oder eingetragene Warenzeichen der jeweiligen Inhaber. Die Wiedergabe von Marken, Produktnamen, Gebrauchsnamen, Handelsnamen, Warenbezeichnungen u.s.w. in diesem Werk berechtigt auch ohne besondere Kennzeichnung nicht zu der Annahme, dass solche Namen im Sinne der Warenzeichen- und Markenschutzgesetzgebung als frei zu betrachten wären und daher von jedermann benutzt werden dürften.

Bibliographic information published by the Deutsche Nationalbibliothek: The Deutsche Nationalbibliothek lists this publication in the Deutsche Nationalbibliografie; detailed bibliographic data are available in the Internet at http://dnb.d-nb.de. Any brand names and product names mentioned in this book are subject to trademark, brand or patent protection and are trademarks or registered trademarks of their respective holders. The use of brand names, product names, common names, trade names, product descriptions etc. even without a particular marking in this works is in no way to be construed to mean that such names may be regarded as unrestricted in respect of trademark and brand protection legislation and could thus be used by anyone.

Coverbild / Cover image: www.ingimage.com

Verlag / Publisher:
LAP LAMBERT Academic Publishing
ist ein Imprint der / is a trademark of
AV Akademikerverlag GmbH & Co. KG
Heinrich-Böcking-Str. 6-8, 66121 Saarbrücken, Deutschland / Germany
Email: info@lap-publishing.com

Herstellung: siehe letzte Seite /
Printed at: see last page
ISBN: 978-3-659-30455-2

Zugl. / Approved by: Islamabad, Air University, Diss., 2012

Acknowledgements

My first and foremost gratitude to "Allah Almighty" for guiding me in good and bad times. For showing me the right path and giving me the courage, potential and ability to complete this project.

There has been a list of people involved in this project to whom I would like to pay my sincere thanks.

First I am obliged to my parents for their patience and in never letting my morale down, my brother who was with me with his full support. To my supervisor Dr. Fida for giving me remarkable suggestions and in providing real-world developments and approach in the implementation of this project. The co-supervisor Mr. Zakwan from the IAA department for letting me get access to the advanced labs and equipment used for research and helping me troubleshoot my deficiencies in this project. The two Jury members Air Commodore Sir Anwar Saeed and Sir Lehaz Ullah Kaka Khel for giving me new directions and technical assistance.

I would also like to thank Mr. Aun for his time and efforts in giving me the detailed information about the Linux kernel and its importance and working. To my friend Shams Ullah for being with me with his time and moral support in all times. And to Mr. Kamran from electronics lab for being cooperative to me.

Table of Contents

List of Figures

Chapter 1: Introduction

This chapter gives the complete introduction of the project and a step by step insight into the process. Real-Time systems are now given particular significance in industry as they are being used to operate machines with great accuracy. Real-Time processing is considered to be one of the important steps towards automation control. Many algorithms exist to do Real-Time tasks but the use of RT-Linux is quite reasonable with respect to cost and efficiency.

1.1 Motivation

Real-Time systems that are being used in industry are quite expensive. My main motivation is to reduce the cost. In order to achieve Real-Time tasking DSP kits and embedded systems are used. They are quite expensive to install in an industry. They require a lot of hardware and need a lot of space.

I am using a personal computer (PC) to achieve Real-Time control. The major advantage of using a PC for Real-Time processing is that I can change the Real-Time behavior as required by just entering instructions from the keyboard. If I were to do the same job using an embedded system then I had to change the entire code written for the controller and burn it again using the controller kit and place it back in the hardware. In case of DSP processors we write a program into the PC and upload it to the DSP kit every time a change is made. Both these processes would be hectic and time consuming and a constant record of the changes should be kept, maintained and observed. But this project will make the Real-Time processing easy, fast and efficient without including any extra hardware and effort.

2

Using Linux operating system was a good and new experience. It was quite challenging to make an operating system meet my needs without including any delay into the process I intend to do. Initially I was not quite familiar with the Real-Time operating systems and this term was new to me but with the passage of time I became quite familiar with the terms and protocols being used to achieve Real-Time control.

1.2 Purpose of the project

My main purpose of this project is to operate dc motor in real time. To run dc motor without any delay so that it can maintain the desired speed I am trying to achieve. According to industry point of view I wanted to automatically control the desired machine in our case dc motor through programming according to any work requirement . This process is also called an industrial automation in Real-Time. One can also perform multitasking through this project.

1.3 Aims and objectives

My aim is to bring the Real-Time system in more advanced and improved form. It is not the first time a Real-Time system is being developed but they were operated using a hardware platform. Making modifications in the process was not an easy task. But my project provides a software platform to easily make modifications in the code.

The system Real-Time output is obtained on parallel port 8 data pins. The response is more accurate and reliable than a serial port that sends a bit by bit data. Parallel port switching is fast and we can achieve more precise control. The world is becoming industrially revolutionized and this little step may contribute towards it.

3

Chapter 2: Background

2.1 Digital Speed Control

2.1.1 Basic concept

Digital speed control is a technique that utilizes switching mode operation to provide any range of speed. In switching mode the computer sends a stream of data consisting of 1's and 0's. A 1 on a port bit corresponds to an ON state and a 0 to an OFF state of that data pin. To achieve digital speed control we need to run a program in order to write that stream of data onto the port. Once the program is running we can attach a motor via an H-bridge to operate it at the desired speed. Varying the duty cycle of the ON vs OFF pulse will give us a variable range of switching and thus a wide range of speeds can be achieved.

2.2 Why DC motor speed control?

2.2.1 DC motor speed control

There are several reasons for the continued popularity of the dc motors. The major application of dc motors is that they can provide a wide variation in speed. Their control is simple and the reaction time is fast. Another reason is that dc power systems are still common in cars, ships and aircrafts, so it makes sense that they will be using dc motors. Even some industries such as textile and paper making industries require speed control of the machines and it can be effectively achieved by the use of the dc motors.

2.3 Components of DC motors

DC motors convert dc electrical energy to mechanical energy. Almost all dc motors have the following components:

1. Rotating part (Armature)

2. Stationary part (Stator)

2.3.1   Armature of a dc motor

Armature is the rotating part of the motor that is made of coils of wire wrapped around the core. And the core has an extended shaft that rotates on bearings. The ends of each coil of wire on the armature terminate at one end of the armature. These termination points are called the commutator. This is the point where the brushes make electrical contact to bring the current from the stationary part to the rotating part of the machine. The current that passes through the armature is called the armature current I_A. This current is responsible for creating the magnetic flux of the rotor.

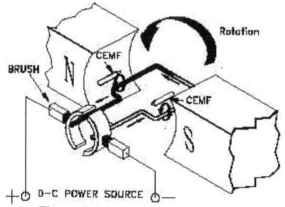

Figure 2.1 Armature of a dc motor

2.3.2   Stator of a dc motor

The stator generates a stationary magnetic field that surrounds the rotor. The field is generated by either permanent magnets or electromagnetic windings. The different types of brushless dc motors are distinguished by the construction of the stator or the way the electromagnetic windings are connected to the power source. The current that flows through the stator is called the field current I_F. This current is responsible for creating the field flux.

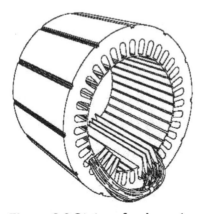

Figure 2.2 Stator of a dc motor

2.4 Types of DC Motors

There are five major types of DC motors in general use:

1. The separately excited dc motor.
2. The shunt dc motor.
3. The permanent magnet dc motor.
4. The series dc motor.
5. The compound dc motor.

2.4.1 The separately excited dc motor

A separately excited dc motor is a motor whose field circuit is supplied from a separate constant-voltage power supply. The Kirchoff's voltage law (KVL) for the armature circuit of the motor is

$$V_T = E_A + I_A R_A$$

The equivalent circuit of a separately excited dc motor is shown.

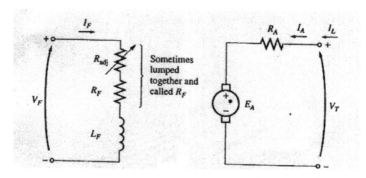

Figure 2.3 Separately excited dc motor equivalent circuit

2.4.2 The shunt dc motor

The shunt dc motor is a motor whose field circuit gets its power directly across the armature terminals of the motor. If the load on the shaft of the shunt motor is increased, then the load torque τ_{load} will exceed the induced torque τ_{ind} in the machine, and the motor will start to slow down. When the motor slows down, its internal generated voltage ($E_A = K\phi\omega \downarrow$) drops, so the armature current in the motor ($I_A = (V_T - E_A \downarrow)/R_A$) increases. As the armature current rises, the induced torque in the motor increases ($\tau_{ind} = K\phi I_A \uparrow$), and finally the induced torque will equal the load torque at a lower mechanical speed of rotation ω. The equivalent circuit of a shunt dc motor is shown.

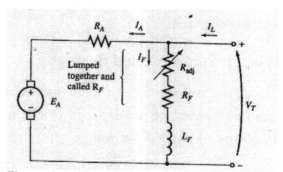

Figure 2.4 Shunt dc motor equivalent circuit

2.4.3 The permanent-magnet dc (PMDC) motor

A permanent-magnet dc (PMDC) motor is a dc motor whose poles are made of permanent magnets, so they do not require an external field circuit. These motors are easy to control and have less field circuit copper losses. As the permanent magnets can not produce a flux as high as an externally supplied field, so a PMDC motor will have a lower induced torque τ_{ind} per ampere of armature current I_A. Moreover, PMDC motors run the risk of demagnetization.

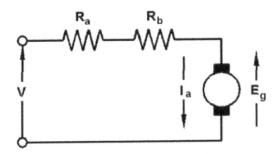

Figure 2.5 PMDC motor equivalent circuit

2.4.4 The series dc motor

A series dc motor is a dc motor whose field windings are connected in series with the armature circuit. So in a series motor, the armature current, the field current, and line current are all the same. The Kirchoff's voltage law equation for this motor is

$$V_T = E_A + I_A(R_A + R_S)$$

The equivalent circuit of a series dc motor is shown on the next page

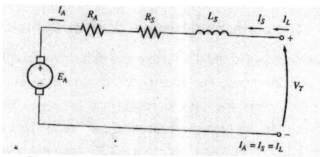

Figure 2.6 Series dc motor equivalent circuit

2.4.5   The compound dc motor

A compound dc motor is a motor with both a shunt and a series field.
The Kirchoff's voltage law for a compound dc motor is

$$V_T = E_A + I_A(R_A + R_S)$$

Equivalent circuit of such a motor is shown below

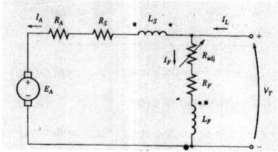

Figure 2.7 Compound dc motor equivalent circuit

2.5 Methods of speed control of DC motors

The most common techniques available for the speed control of a dc
motors:

1. Change the field resistance R_F

2. Change the armature voltage V_A

3. Change the armature resistance R_A

2.5.1   Field resistance R_F control

When we increase the field resistance R_F following happens:

1. Increasing R_F causes $I_F = V_T / R_F \uparrow$ to decrease.
2. Decreasing I_F decreases ϕ.
3. Decreasing ϕ lowers $E_A = K\phi \downarrow \omega$.
4. Decreasing E_A increases $I_A = (V_T - E_A \downarrow)/R_A$.
5. Increasing I_A increases $\tau_{ind} = K\phi \downarrow I_A \Uparrow$, with the change in I_A dominant over the change in ϕ.
6. Increasing τ_{ind} makes $\tau_{ind} \succ \tau_{load}$, and the speed ω increases.
7. Increasing ω increases $E_A = K\phi\omega \uparrow$ again.
8. Increasing E_A decreases I_A.
9. Decreasing I_A decreases τ_{ind} until $\tau_{ind} = \tau_{load}$ at a higher speed ω.

2.5.2   Armature voltage V_A control

When we increase the armature voltage V_A following happens:

1. An increase in V_A increases $I_A = (V_A \uparrow - E_A)/R_A$.

2. Increasing I_A increases $\tau_{ind} = K\phi I_A \uparrow$.

3. Increasing τ_{ind} makes $\tau_{ind} \succ \tau_{load}$ increasing ω.

4. Increasing ω increases $E_A = K\phi\omega \uparrow$.

5. Increasing E_A decreases $I_A = (V_A \uparrow - E_A)/R_A$.

6. Decreasing I_A decreases τ_{ind} until $\tau_{ind} = \tau_{load}$ at a higher ω.

2.5.3   Armature resistance R_A control

The speed of series dc motors also be controlled by the insertion of a series resistor into the motor circuit. But this technique is very wasteful of power and is used only for the start-up of some motors. This method is found only in applications in which the motor spends almost all its time operating at full speed or in applications too inexpensive to justify a better form of speed control, than this inefficient method.

2.6 What is Real-Time system?

2.6.1 Real-Time system

A 'real-time system' is one in which the correctness of the computations not only depends upon the logical correctness of the computation but also upon the time at which the result is produced. If the timing constraints of the system are not met, system failure is said to have occurred. In other words, the system must be deterministic to guarantee timing behavior in the face of varying loads. In addition to bringing determinism to interrupt processing, task scheduling that supports periodic intervals is also needed for real-time processing.

2.6.2 Types of Real-Time systems

Real-Time systems are classified into:

1. Hard Real-Time systems.
2. Soft Real-Time systems.

2.6.2.1 Hard Real-Time systems

An operating system that can support the desired deadlines of the real-time tasks (even under worst-case processing loads) is called a 'hard real-time' system. But hard real-time support isn't necessary in all cases. In hard real-time systems are those in which missing a deadline can have a catastrophic result such as opening an airbag too late.

2.6.2.2   Soft Real-Time system

If an operating system can support the deadlines on average, it's called a 'soft real-time' system. Soft real-time systems can miss deadlines without the overall system failing such as losing a frame of a video.

2.7 What is PWM technique?

2.7.1   PWM technique

A simplest method to control the rotation speed of a DC motor is to control its driving voltage. The higher the voltage is the higher speed the motor tries to reach. In many applications a simple voltage regulation would cause lots of power loss on control circuit, so a pulse width modulation method (PWM) is used in many DC motor controlling applications. In the basic Pulse Width Modulation (PWM) method, the operating power to the motors is turned on and off to modulate the current to the motor. The ratio of "on" time to "off" time is what determines the speed of the motor. When doing PWM controlling, keep in mind that a motor is a low pass device. The reason is that a motor is mainly a large inductor. It is not capable of passing high frequency energy, and hence will not perform well using high frequencies. Reasonably low frequencies are required, and then PWM techniques will work. A higher PWM frequency will work fine if you insert a large capacitor across the motor. The idea that a lower frequency PWM works better simply reflects that the "on" cycle needs to be pretty wide before the motor will draw any current (because of motor inductance).

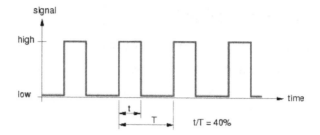

Figure 2.8 PWM signal

Chapter 3: Project Description

3.1 General Idea

* Install 'Linux' operating system on some desktop PC preferably "Red Hat Linux v3.5" or higher.

* Run and configure 'Real Time Application Interface' (RTAI) kernel (any stable version) on the same PC on which you installed 'Linux".

* Reboot the system then write the code given in appendix 'A' in some text file with extension '.c' and compile it.

* Connect the motor drive circuitry (H-Bridge) with the parallel port of the PC along with the dc motor attached to it.

* Run the code on the PC, enter your desired duty cycle and the motor will run on the required speed.

3.2 Block Diagram

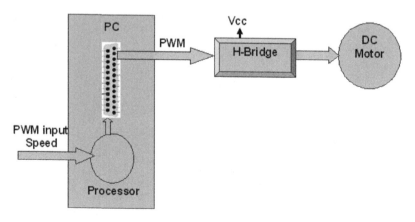

Figure 3.1 Block Diagram

3.3 Algorithm for speed control

The whole process is simple and easy. All one needs is some PC with desired ingredients and a keyboard (mouse as well). Open the Linux terminal, get inside the RTAI patch, run the compiled code, type in the duty cycle. And the result is the control of the dc motor in your hands. If you need some higher/lower speed then just terminate the program by pressing any key, re-run the compiled code, re-enter some different duty cycle and press enter. You will notice the change in the motor speed. It is all that simple. But be aware that on lower duty cycles or low speeds the motor run would be jerky. The best variation of speed can be achieved for duty cycles in the range 50-100.

3.4 Flow diagram of process

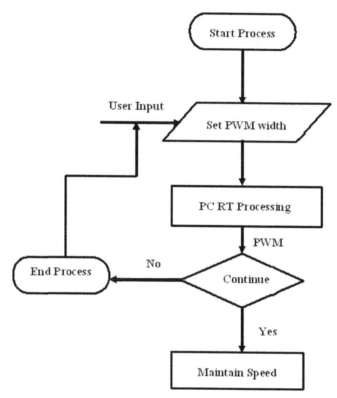

Figure 3.2 Flow Diagram

Chapter 4: Hardware Selection

4.1 Transistor Selection

4.1.1   Tip transistor advantage over typical Bipolar Junction Transistors (BJT's)

Tip transistor is a power BJT that uses Darlington pair configuration. Darlington pair uses two transistors to provide a high gain. The collector terminals of the two transistors are connected together to provide common biasing to both the transistors. The base of the first transistor is the active base of the pair, while the emitter of the first transistor connects to the base of the second transistor. This lets the collector current of the first transistor to flow directly through the base of the second one and thus providing it a high current gain. The emitter of the second transistor is used as a common-emitter of the pair. The forward voltage drop from the base to the emitter of the Darlington's is approximately two times the forward voltage drop of a single transistor. Major characteristic of a Darlington is its high gain ranging from 100 to 1000. It is a beta multiplier and therefore has a higher emitter current capability. It has higher current capability as compared to a regular transistor. It is an ideal choice for providing higher current handling, higher power output and efficient heat dissipation than a normal BJT.

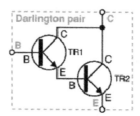

Figure 4.1 Tip transistor internal circuit

4.2 Voltage Regulation

4.2.1 Voltage Regulator LM7805

A voltage regulator is to take a variable input voltage and produce a constant regulated output voltage. LM7805 is a 5V voltage regulator used to ensure that no more than 5V is delivered to the circuit. A regulator functions as a diode to clamp the output voltage at 5V dc regardless of the input voltage. The excess voltage is converted to heat and is dissipated through the metal body of the regulator. To dissipate heat we must use heat sinks to protect the device from damage of over heating. The regulator has three legs the input, ground and output of 5V. It is an efficient device and does not draw much power from the circuit.

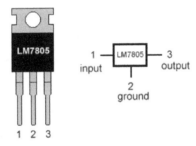

Figure 4.2 Voltage regulator and its pin configuration

4.3 DC Motor selection

4.3.1 Permanent Magnet DC (PMDC) motor

Direct Current (DC) motor is an actuator that converts electrical power into mechanical power. A permanent magnet DC (PMDC) motor consists of a permanent magnet stator and armature windings in the rotor. The armature winding is supplied with a DC voltage that causes a DC current

21

to flow in the windings. Interaction between the magnetic field produced by the armature current and that of the permanent magnet stator causes the rotor to rotate. These two fields result in a torque which rotates the rotor. As the rotor turns, the current in the windings is commutated to produce a continuous torque output on the shaft.

4.3.2   Advantages of PMDC motor

There are several typical advantages of a PM motor. A PMDC motors provides very precise control and speed regulation over a wide range. There is no separate field excitation needed. PM motors are physically smaller in overall size and lighter for a given power rating. Since the motor's field, created by the permanent magnet, is constant, the relationship between torque and speed is therefore linear. A PM motor can provide relatively high torque at low speeds and permanent magnet field provides some inherent self-braking when power to the motor is turned off. They are portable and have linear speed-torque characteristics and are adaptable to various types of control methods.

4.3.3 Exciting a PMDC motor

The circuit below includes the resistance R_a and inductance L_a of the motor's armature windings. When a DC supply voltage V_a is applied to the armature, current I_a flows in the armature and the torque τ_{ind} is induced. Due to the movement of the armature in the magnetic field a back-emf voltage e_{ind} is generated that is proportional to the speed of the motor ω.

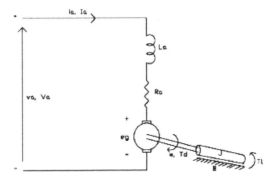

Figure 4.3 PMDC motor Equivalent Circuit Diagram

4.3.4 Finding the unknown DC motor parameters

To drive a motor one must know the relationship between motor torque and current, and the relationship between motor speed and generated e.m.f. Normally, when we obtain a DC motor, we will not know all the information necessary to characterize our motor in terms of unknown parameters. These unknown parameters are motor constant, internal resistance of the motor, magnetic field etc.

One way to characterize our motor is to look on the package of the motor, and obtain all the information one can get from there. Then use the given information in equations given below to calculate the values of K, I_a and R_a or other parameters.

$$e_{ind} = K\Phi\omega \qquad \tau_{ind} = K\Phi I_a \qquad E_a = V_t - I_a R_a$$

4.3.5 Direction of rotation and speed control of a PMDC motor

Sometimes the rotation direction needs to be changed. In normal permanent magnet motors, this rotation is changed by changing the polarity of operating power for example by inter-changing the power terminals going to power supply. This direction changing is typically implemented using a circuit called an H bridge.

A simplest method to control the rotation speed of a DC motor is to control its driving voltage. The higher the voltage is the higher speed the motor tries to reach. In many applications a simple voltage regulation would cause lots of power loss on control circuit, so a pulse width modulation method (PWM) is used. In the basic Pulse Width Modulation (PWM) method, the operating power to the motors is turned on and off to modulate the current to the motor. The ratio of "on" time to "off" time is what determines the speed of the motor.

4.3.6 Horse power output of the motor

Horse power is the measure of the rate at which the work is done. Horse power is the power output of the motor at the base speed. The base speed gives an indication of how fast the motor will run with rated armature voltage and rated load current at rated flux. The motor nameplate provides specific critical information including the horse power, RPM and rated voltage along with other parameters. Horse Power (HP) is given as:

$$HP = (\tau * RPM) / 5250 = (V * I * \eta) / 746$$

It shows that a decrease in speed (RPM) produces a proportional decrease in Horse power (HP) where one HP equals 746 watts.

4.4 Calculating the transfer function of the motor

A motor is a machine that takes in electrical input and gives out mechanical output. Now considering the input is the motor voltage and the output is the motor velocity. Here is how we calculate the transfer function of our motor on Matlab Simulink. Where:

J= Moment of Inertia of the rotor

B= Damping ration of the mechanical system

K= Electromotive force constant

R= Electric resistance

L= Electric inductance

     V= Input source votage

θ = Ouptput position of shaft

Draw the following function in Matlab Simulink and then simulate:

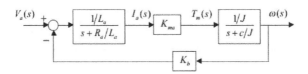

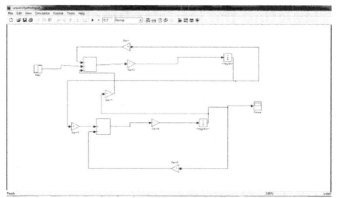

Figure 4.4 Transfer function calculation on Matlab Simulink

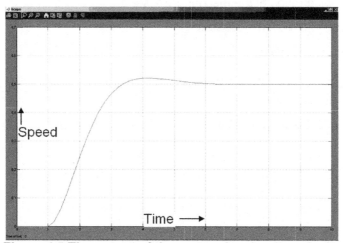

Figure 4.5 The output of the motor showing its velocity

Initially when the voltage is applied the motor has a high starting torque. The motor tries to build speed and its torque decreases. When the voltage reaches the rated voltage the motor achieves the rated speed. After this the motor runs on the steady speed as long as the voltage and the load is kept unchanged. The transfer function is second order and is given by:

$$G(s) = \frac{K_\tau}{JR_m s^2 + (BR_m + K_\tau K_m)s}$$

26

4.5 Parallel port for generating PWM

On PCs, the parallel port uses a 25-pin connector that is used to connect printers, computers and other devices that need relatively high bandwidth. When a PC sends data to a printer or other device using a parallel port, it sends 8 bits of data (1 byte) at a time. These 8 bits are transmitted parallel to each other. The standard parallel port is capable of sending 50 to 100 kilobytes of data per second.

Let's take a closer look at what each pin does when used with a printer:

- Pin 1 carries the strobe signal. It maintains a level of between 2.8 and 5 volts, but drops below 0.5 volts whenever the computer sends a byte of data. This drop in voltage tells the printer that data is being sent.
- Pins 2 through 9 are used to carry data. To indicate that a bit has a value of 1, a charge of 5 volts is sent through the correct pin. No charge on a pin indicates a value of 0. This is a simple but highly effective way to transmit digital information over an analog cable in real-time.
- Pin 10 sends the acknowledge signal from the printer to the computer. Like Pin 1, it maintains a charge and drops the voltage below 0.5 volts to let the computer know that the data was received.
- If the printer is busy, it will charge Pin 11. Then, it will drop the voltage below 0.5 volts to let the computer know it is ready to receive more data.

27

- The printer lets the computer know if it is out of paper by sending a charge on Pin 12.

- As long as the computer is receiving a charge on Pin 13, it knows that the device is online.

- The computer sends an auto feed signal to the printer through Pin 14 using a 5-volt charge.

- If the printer has any problems, it drops the voltage to less than 0.5 volts on Pin 15 to let the computer know that there is an error.

- Whenever a new print job is ready, the computer drops the charge on Pin 16 to initialize the printer.

- Pin 17 is used by the computer to remotely take the printer offline. This is accomplished by sending a charge to the printer and maintaining it as long as you want the printer offline.

- Pins 18-25 are grounds and are used as a reference signal for the low (below 0.5 volts) charge.

With each byte the parallel port sends out, a handshaking signal is also sent so that the printer can latch the byte.

Figure 4.6 Standard 25 pin Parallel port female connector

4.6 Shaft Encoder for closed loop control

4.6.1   Closed loop control

To obtain accurate results with extremely short response time, then we need a closed loop control. In order to build a closed loop controller, we need to gain information about the rotation of the shaft like the number of revolutions executed per second and the precise angle of the shaft. This source of information about the rotations of the shaft of the motor is provided by a shaft encoder. A 'Shaft Encoder' is a devise that will translate the rotation of the shaft into electrical signals that can be communicated to the controller. In other words, a closed loop controller will regulate the power delivered to the motor to reach the required velocity. If the motor is to turn faster than the required velocity, then controller will deliver more power to the motor. Control of the electrical power delivered to the motor is usually done by Pulse Width Modulation.

4.6.2   Shaft encoder operation

An encoder disk is one of the main parts of a shaft encoder. A U-shaped photo-couple made of an Infra-Red sender and a receiver is positioned in a way so that the beam of infrared light passes through one of the small openings in the encoder disk. The photo-couple is composed of a LED and a photo transistor. The encoder disk is connected to the back-shaft of the motor, so that both the shaft and the encoder disk rotates at the same R.P.M. The rotation of the motor causes the beam of light to be periodically intercepted by the solid parts of the encoder disk creating a sequence of pulses of light that are translated by the photo couple's receiver into pulses of electricity. Those pulses of electricity contain all the information we need to implement a closed loop control. The frequency of those pulses is directly proportional the speed of rotation of the shaft RPM and the number of those pulses correspond to the angular displacement of the shaft. The more the number of holes in an encoder disk, the higher will be the resolution and thus the slightest angular displacement that can be detected.

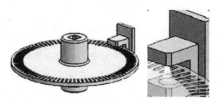

Figure 4.7 Shaft encoder disk along with IR sensor

Chapter 5: Software Installation

5.1 Introduction to Linux operating system

Linux is a very powerful operating system and it is gradually becoming popular throughout the world. Rarely does it frees up or slow down. There is no need to reboot the Linux system to maintain performance levels. Linux provides high performance. It is a very flexible multitasking operating system and can handle many things at the same time. Linux can be used for high performance server applications, desktop applications, and embedded systems. One can install only the needed components for a particular use. Linux is an Open source operating systems. You can easily get the source code for Linux and edit it to develop your personal operating system. Today, Linux is widely used for both basic home and office uses. It is the main operating system used for high performance applications and in web servers. Linux has made a high impact in this world.

5.2 What is a Linux Kernel?

A Linux kernel is a set of instructions already written and defined into the machine language that instructs a computer to perform the required task when a specified condition is full filled. It is a free and open source available on the internet and can be modified according to our requirement. There are many types of kernels developed so far but the most significant is the Real-Time (RT) kernel. This kernel provides 'preemptive scheduling' i.e. an execution of a current process in user mode would be interrupted if a higher priority process entered in task running state. This means that if a low-priority process makes a system call, a high-priority process does not wait and can gain access to the CPU immediately even if the process is in the middle of a system call.

Thus in RT kernel, one can get soft real-time performance through a simple kernel configuration.

5.3 Installing Real-Time Application Interface (RTAI) kernel on Linux

5.3.1   System Requirements

- Intel Processor, Pentium III or higher at 500 MHz or higher.
- Disk capacity of 9 GB including the 1.1 GB free space for kernel compilation.
- RAM memory of 256 MB. 128 MB of RAM may also work but the compilation process and OS installation is very slow.
- CD ROM Drive of 50X speed. Slower drives will slow down the operating system setup procedure.

5.3.2   Software Requirements

- Operating system, Red Hat® Enterprise Linux® version 4 or higher.
- Compiler, GCC v3.4 and G++ v3.4. To check currently installed GCC and G++ version simply type this command in the terminal:

$ gcc --version
$ g++ --version

This will print the versions of these programs. Then type the following:

sudo –s

This will allow you to get super user privileges throughout the installation process. Now, to get the above mentioned packages type the following:

 # apt-get install gcc-3.4 g++-3.4 make

One may already have the higher versions installed in the system. So when we type the **apt-get** command the above, the packet manager installs the lower version in /usr/bin, but lets the newest version available and running as default. To use the recommended versions, we need to specify it at the compilation stage.

- Basic kernel configuration menu, libncurses5-dev. This package installs the needed curses libraries that we will need to run the kernel configuration menu. Simply type:

 # apt-get install libncurses5-dev

- Module loader, module-init-tools. These tools will be needed to load kernel .ko modules, such as the rtai_hal.ko. To get it type the following:

 # apt-get install module-init-tools

- RTAI version 3.5. This is the stable RTAI version until now, and has patches to the recent 2.6.19 kernel version. To get this software, first enter the source directory address:

 # cd /usr/src

Now get the RTAI tar file to this same folder.

#wget--no-check-certificate https://www.rtai.org/RTAI/rtai-3.5.tar.bz2

And unpack the tar file.

 # tar xvf rtai-3.5.tar.bz2

This creates the folder rtai-3.5.tar.bz2 on your /usr/src directory.

5.3.3 Applying RTAI Patch

Now we have the necessary tools and are able to start working. The RTAI kernel patches are applied to our operating system, so it can later support the RTAI kernel layer. To do so, enter the folder of your kernel sources:

 # cd /usr/src/linux-2.6.19

And apply the patch:

 # patch -p1 -b < ../rtai-3.5/base/arch/i386/patches/hal-linux-2.6.19-i386-1.7-01.patch

One can also use the tab key to auto complete the commands on the terminal. After this, the process will show that the patch is being applied.

5.3.4 Configuring the Kernel

For this step one needs to recognize its hardware over which the kernel is being installed. It is straight forward to configure a desktop pc. For configuring it on laptops first see the specific laptop modules that will support the software.

First save the current kernel configuration file. Linux distros stores a copy of this file, which will help us with a starting configuration. So enter in the kernel source folder by typing:

 # cd /usr/src/linux-2.6.19

and make a copy of the existing configuration file to the root folder of your kernel source:

 # cp /boot/config-2.6.19-generic .config

Now finally run the kernel configuration menu.

 # make menuconfig CC=/usr/bin/gcc-3.4 CXX=/usr/bin/g++-3.4

The curses base menu will show up. Now load the complete configuration file that we saved as .config by choosing the 'Load an alternate configuration file' option and typing .config if it is not already as default. At this point load the kernel configuration of the default Linux installation which is running on your system.

5.3.4.1 The configuration

On the kernel configuration menu, do the following:

- Code maturity level options ->

    - Nothing selected

- General setup ->

  - [*] Support for paging of anonymous memory (swap), support for swap (virtual memory).
  - [*] System V IPC, allows inter-process communication.
  - BSD process accounting, allows to obtain user application information's.

- Loadable module support ->

  - !![*] Enable loadable module support, allows to load modules to the kernel with the loading tools.
  - [*] Module unloading.
  - [*] Source checksum for all modules.
  - [*] Automatic kernel module loading.

- Block layer ->

  -Nothing selected

- Processor type and features ->

  - [*] Generic x86 support, better kernel performance on x86 architecture CPU's.
  - [*] Preempt the big kernel lock, reduces latency of the kernel on desktop computers.
  - [*] Interrupt pipeline, prevents data disturbances.

- !![] Local APIC support on processors. It must be deactivated or the error:
- RTAI [hal]: ERROR, Local APIC Configured But Not Available/Enabled will show when running RTAI apps.
- [*] Math emulation, emulates co-processor for loading point operations on old CPU's.
- [*] MTRR support.
- !![] Use register arguments, this must be deactivated.
- [*] Compact VDSO support.

- Power management options ->

  - [*] Legacy management debug support
  - ACPI support ->

    - o [*] ACPI support, advanced configuration and power interface support.
    - o [M] Button
    - o [M] Video
    - o [M] Fan
    - o [M] Processor
    - o [M] Thermal Zone

  - CPU frequency scaling ->

    - o CPU frequency scaling, allows to change the clock frequency of the CPU on the fly.

o Relaxed speedstep capability checks, does not perform all checks for speed up.

- Bus options ->

  - [*] PCI support

- Executable file formats ->

  - Kernel support for ELD binaries.

- Networking ->

  - Networking options ->

    o [*] Packet socket: Mapped IO, speed up communications.
    o [*] Unix domain sockets, supports UNIX sockets.
    o [*] TCP/IP networking, all these options will be marked.
    o [*] Network packet filtering.
    o [*] QoS and fair queuing.

- Device drivers ->

  - Generic driver options ->

    o [*] Prevent firmware from being built.
    o [*] User space firmware loading support.

- Memory technology devices (MTD) ->

    o Write support for NFTL.

- Plug and Play support ->

    o [*] Plug and Play support

The rest of the options use the default configuration of the running kernel.

- File systems, default configuration used.

- Kernel Hacking ->

    - !![] Compile the kernel with frame pointers, must be deactivated. The rest of the options use the default configuration of the running kernel.

- Security options, default configuration used.

- Cryptographic options, default configuration used.

- Library routines, default configuration used.

This process requires that you know your hardware if you want to optimize your kernel's performance. The above configuration would work in any desktop equipped with an x86 CPU

5.3.5   Compiling the Kernel

After configuring the kernel comes the compilation process. Some required software packages are:

 # apt-get install kernel-package fakeroot

Now run the following commands to clean and compile the kernel.

 # make-kpkg clean

 # fakeroot make-kpkg --initrd --app\end-to-version=-rtai \kernel_image kernel_headers

This process may take hours to complete.

When the process is finished, a line is displayed 'echo done'. Now install the two *.deb packages that are created in the /usr/src folder by the compilation process.

 # cd /usr/src

 # dpkg -i *.deb

After the dpkg program installation is finished, a new entry in the menu.lst of the grub will be added. So if we reboot the system now, this new entry will boot the Linux system with our new kernel.

5.3.6   Configuring RTAI

After completing the above steps, boot up you new kernel. Now, enter the rtai-3.8 folder and create a new one for build:

$ sudo –s

 # cd /usr/src/rtai-3.5

 # mkdir build

 # cd build

Now configure the RTAI:

 # make -f ../makefile CC=/usr/bin/gcc-3.4 CXX=/usr/bin/g++-3.4

Verify the following options in the menu that will show up:

- General -> Installation directory, leave the default as usr/realtime
- General -> Linux build tree, the patch to the configured kernel /usr/src/linux-2.6.9.

All is set, now exit and reply YES to save the configuration. Now, install RTAI:

 # make install

Now reboot the computer and boot our new kernel with RTAI installed in it.

Chapter 6: Hardware and Software Implementation

6.1 Hardware Implementation

6.1.1   Motor driving circuit

An H-bridge circuit drives a dc motor. It has two logic level inputs A and B and two logic level outputs A and B. If input A is brought high, output A goes high and output B goes low. The motor rotates in one direction. If input B is driven, the opposite happens and the motor runs in the opposite direction. If both inputs are low, the motor is not driven and can freely "coast", and the circuit consumes no power. If both inputs are brought high, the motor is shorted and braking occurs. The H-bridge circuit schematic is shown below.

To do PWM speed control we need to provide PWM pulses to the circuit. The PWM is applied to one input or the other based on the direction desired and the other input is held low. Depending on the frequency of the PWM and desired reaction of the motor.

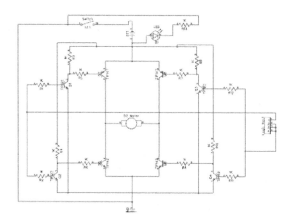

Figure 6.1 H-bridge circuit

6.1.1.1   Bridge stage

The first and the most important stage in an H-bridge circuit is the bridge stage. It not only delivers power to the motor but also instructs the motor that in what direction it is desired to rotate. It has four high power transistors, two are Tip 142 and two are Tip 147. Tip 142 is NPN and Tip 147 is PNP. At one time one NPN and one PNP is turned on in such a way that current flows through the motor in one direction suppose and makes the motor run suppose counter clockwise. While if the other pair is turned on while keeping the first pair of transistors off then the current flows in the opposite direction and the motor rotates in clockwise direction.

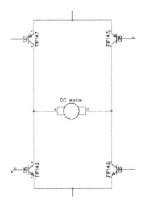

Figure 6.2 Bridge stage

6.1.1.2   Outer stage

The circuit outside the bridge stage supports the inner stage. It consists of transistors and resistors. This stage provides biasing to the next inner stage described above. It uses Tip- 122 transistors that are an NPN power Darlington transistor. It provides current biasing to the Tip 122 and Tip 147 transistors. The collector current that flows through the Tip

122 transistor also flows through the base of the inner stage transistors through a voltage divider and causes them to turn on.

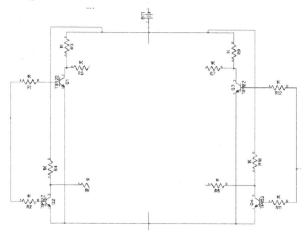

Figure 6.3 Outer stage

6.1.2 Schematic of H-bridge

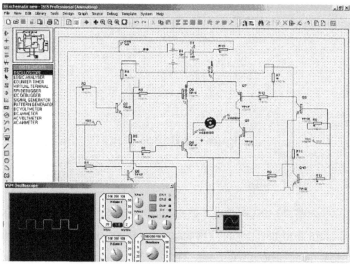

Figure 6.4 Schematic of H-bridge

6.1.3 ADC and DAC conversions

Although this part is not actually being used in out project but I have used these converters just in case I needed to do the conversion if required.

6.1.3.1   Analog to Digital conversion

An A to D converter takes in the analog input and gives an 8 bit digital output. It compares the input signal with a 5V reference signal and based on this comparison gives the digital word. This digital output can then be made input into any digital computer. We can do the required manipulations on this signal and code it according to our requirement.

6.1.3.2 Digital to Analog conversion

A D to A converter takes in the digital word and converts it into an analog signal. The output signal can be inverting or non-inverting. We may use this converter to convert the digital output coming form the parallel port into an analog signal that can then be fed to the H-bridge circuit.

6.1.4   Optocoupler

Optocoupler consists of an LED and a photo transistor. The purpose of an opto coupler is to separate two parts of a circuit. This reduces interference and thus reduces the chances of error encountered in the circuit. It also provides protection to the circuit in case of system failure. Optocoupler needs two supplies in order to function. I am providing one supply from the parallel port and the other supply is given to bias the photo transistor. The output is obtained from the collector of the photo transistor.

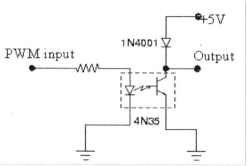

Figure 6.5 Optocoupler circuit

6.2 Software Implementation

6.2.1 Vim editor for coding

Vim is a powerful text editor in Linux that can be used to write, store and compile a program in Linux. Vim editor includes all the commands necessary for a Hardware interface.

There are three modes of vim editor:

1. Command mode
2. Insert mode
3. Save and Exit mode

6.2.1.1 How to insert, save and exit a code in vim editor

Step 1: Type **vim filename.c**

This creates a file named filename with extension of C code.

Step 2: Type **i**

To switch to insert mode to enter text in the file.

Step 3: Enter text

Write your program code in it.

Step 4: Press **Esc** key

To switch back to command mode.

Step 5: Type **::wq**

To save text in the file and exit vim.

6.2.1.2 To compile and run the code

First enter the Real-Time mode by writing the following command in the "terminal window":

cd rtai-3.5

To compile the program write:

g++ filename.c –o codefile

This notifies if the program contained any errors, run the program with the command:

./codefile

6.2.2   Getting Speed from the motor

The speed is manually obtained from a 'tachometer'. It is a device that measures the speed of the motor. It directs a laser beam onto the rotating reflecting surface and notices its reflection. Based on this it can find the number of revolutions executed by the motor. Before we can start taking the calculations we need to set the reference point on the motor shaft. This can be done by pasting a reflecting tape on the shaft. Start the motor and hold the tachometer gently upon the shaft of the motor. Press the button on the device and it will give you a stable RPM of the motor after few seconds.

6.2.3   PC speed processing

The PC will get the duty cycle input. And will convert these pulses to the ON and OFF duration of the time period T written in the code. E.G if the ON time is 't' times T, then the off time will be '1-t' times T so that their sum equals the total period of time T. The PC will write 1 on the parallel port in the ON duration and 0 in the OFF duration. The sequence of these 0's and 1's and their duration and in turn the motor speed can be altered by entering the duty cycle input into the computer.

Chapter 7: Conclusion

7.1 Conclusion

Our project with the grace of Allah Almighty proved to be a success. It is a leap towards industrial automation. Towards achieving Real-Time control and perhaps towards the end of an era that used embedded system and DSP kits to serve the same purpose and get the job done. The algorithm I used was easy and the whole process was user friendly. One should only enter the data with the keyboard and the system would do the rest of the job. The chances of error were also dilute.

Almost any one with a PC installed with a Real-Time Operating System (RTOS) can perform any of the jobs requiring the Real-Time computing. So this kind of system has been easily available at some cheap price. Now you just need to prepare the software, run it and connect the required hardware. The PC processor will assist you in getting the work done.

Chapter 8: Applications and Future Enhancements

8.1 Applications

There have been many applications in which Real-Time computing is being used to control certain processes that need event scheduling and time constraint.

Here is a list of some common Real-Time examples in our daily life:

- A standard kernel can be used to control an elevator to give hardware interrupts to the micro switches at an RT weight.

- Anti-locking brakes on a car are also and example of a Real-Time computing system, where time is the constraint in which the brakes must be released to prevent the wheels from locking.

- Deploying of an air bag in a car in case of an accident is a Real-Time process in which the response time of the air bag should be in milliseconds to prevent fatal injury.

- In medical terms the equipment of ECG and MFT give Real-Time activity of the heart and brain of the patient under treatment.

- The transaction of amount from a credit card is also in Real-Time in which the machine checks your credit card and within seconds you are approved.

- Fly-by wire systems in modern airplanes uses Real-Time processing of the electrical signals to control the position of the flaps and rudder.

- The textile, paper making and printing press uses motors to rotate the rollers at an equal pace and get the job done without and delay, otherwise the process would be faulty and waste of material and precious time.

8.2 Future Enhancements

This project can be used in a number of applications with some little modification. The hardware that I interfaced with the computer was a dc motor. One can use the same motor and can attach an elevator as a load. This can be used to lift the passengers. It can be used in auto flight for aero planes, dc motors can control the flaps at the wings that in turn maintains the feeded direction and height. It can be used in industrial automation to reduce the need for human work in the production of goods. It can be used in electronic trains to automatically control their speeds in accordance with the sharpness of a turn. Steel mills, paper making machines use motors that can be programmed at any speed and thus eliminate the need to buy multiple motors that run at different speeds.

Appendices 'A'

Source Code

Source Code

The following code generates the required duty cycle on the parallel port that will be transferred to hardware for the motor speed control.

```c
#include <stdio.h> /* for printf() and scanf() */

#include <iostream> /* for input output operations */

#include <stdlib.h> /*used to communicate with PC */

#include <unistd.h> /* needed for ioperm */

#include <sys/io.h> /* for outb() and inb() */

#include <sys/types.h> /*gain access to LPT */

#include <fcntl.h> /* file control options */

#define BASEPORT 0x378 /* lp1 */

int main() {

        char c;
        int n, tem;
        int on,off;
        int duty;

        printf("Enter Duty Cycle 1-100\n");
        scanf("%d",&duty);

        int T=500;
        on=T*duty;
        off=(100-duty)*T;

        printf("Hit any key to Stop\n");

        if (ioperm(BASEPORT, 3, 1)) {perror("ioperm"); exit(1);}

        tem = fcntl(0, F_GETFL, 0);
        fcntl (0, F_SETFL, (tem | O_NDELAY));
```

```
        while (1) {
        n = read(0, &c, 1);
        if (n > 0) break;

        outb(255, BASEPORT);
        usleep(on);

        outb(0, BASEPORT);
        usleep(off);
    }

    fcntl(0, F_SETFL, tem);
    outb(0, BASEPORT);

    if (ioperm(BASEPORT, 3, 0)) {perror("ioperm"); exit(1);}

    exit(0);
}
```

References

[1]   Balagurusamy, "Programming in ANSI C", 5th Edition, Mc Graw Hill, 2011.

[2]   Balagurusamy, 'Constants, Variables and Data Types' in "Programming in ANSI C", Mc Graw Hill, 2011, pp. 23-49.

[3]   Chapman, "Electric Machinery Fundamentals", 3rd Edition, Mc Graw Hill, 1999.

[4]   Chapman, 'DC Motors and Generators' in "Electric Machinery Fundamentals", Mc Graw Hill, 1999, pp. 506-546.

[5]   H.Rashid, "Power Electronics Habdbook, Circuits, Devices and Applications", 3rd Edition, Pearson, 2011.

[6]   H.Rashid, 'DC Motor Drives' in "Power Electronics Handbook, Circuits, Devices and Applications", Pearson, 2011, pp. 917-919, 943-946.

[7]   Joao Monteiro, "RTAI Installation Complete Guide", Version 1-2, February 27, 2008.

[8]   Keith Shortridge, "A Guide to Installing RTAI Linux", Anglo-Australian observatory, Novermber 12th 2004.

[9]   RTAI 3.4 User Manual, October 2006.

[10] http://www.scribd.com/doc/

[11] http://ubuntuforums.org/

[12] http://cplusplus.com/reference/

Figure References

[1]   Figure 2.1;
http://nuclearpowertraining.tpub.com/h1011v2/img/h1011v2_109_1.jpg

[2]   Figure 2.2; http://threephaseelectricmotor.com/wp-content/uploads/2012/03/stator-induction-motor.jpg

[3]   Figure 2.3, Figure 2.4, Figure 2.6, Figure 2.7;
"Electric Machinery Fundamentels" by 'Stephen J.Chapman', 4th Edt.

[4]   Figure 2.5;
http://www.globalspec.com/RefArticleImages/2E7925ABC41B09A0B5EEC3EC31C13BF3_9_1_1_9_1_1-IMGS-14.jpg

[6]   Figure 2.8; http://www.pabr.org/pxarc/1.1/doc/hbridge_signal.gif

[7]   Figure 4.1; http://www.kpsec.freeuk.com/images/darlingt.gif

[8]   Figure 4.2;
http://georobot.files.wordpress.com/2012/09/lm7805-pinout-diagram.gif

[9]   Figure 4.3; http://www.wiringdiagrams21.com/wp-content/uploads/2009/09/Permanent-Magnet-DC-Motor-Equivalent-Circuit-Diagram.png

[10]   Figure 4.6; http://2.bp.blogspot.com/-RSVGM9fJSBQ/Tk-yHnrC7BI/AAAAAAAAfU/9by6cOS5U1o/s1600/Parallel-Port.jpg

[11]   Figure 4.7;
http://www.vexforum.com/wiki/images/2/22/Optical_Shaft_Encoder_Figure_2.jpg

[12]   Figure 6.5;
http://www.dca.fee.unicamp.br/courses/EA079/2s2005/projetos/ventilador/acionador_isolado.gif